THE NO-NEUTRON ATOM
Alternative Atom Models

Dean LeRoy Sinclair (BA, MS, PhD}

Copyright, 2017, Dean L. Sinclair

Book Summary

This book is intended as an introduction to the changes in te way that atoms may be viewed in the light of the Oscillators in a Substance Model of Existence* which adds new insights and new model forms which give answers to previous mysteries by supplementing, an bd sometimes supplanting , existing models of the atom. The most obvious supplant is that ideas from the Oscillators in a Substance Model suggest that neutrons, as such, do not exist in atomic nuclei. The view of existence defined by the title of the model, Oscillators in a Substance; that is, the idea that Existence is within a Substance at , or near, its triple point, which is organized into, and/or, by, oscillators gives a very different, yet apparently consistent and logical view of what atoms may be and, indeed, what our universe may be.

<u>*Sinclair, Dean; Oscillators in a Substance Model of Existence': A Physicist's Grail and an Alexander's Sword</u>, Amazon Createspace, 2015.

Background: The Oscillators in a Substance Model, in Brief.

The Oscillators in a Substance Model is the "Twilight Years Hobby" of a currently 85 year old former science teacher, who at the age of 72, in 2004, realized that Einstein's Relativity Work was not actually theory; but, a mathematical description of information change with viewpoint and motion of transmitter and receiver in a particular "Perceptual-Universe-Defined-by-a-Maximum-Velocity-of-INformation-Transfer."
This generalization of Einstein's work led to a new view of the significance of the Speed of Light as an Average-Velocity-of-Information-Carriers-OVer-any-Significant-Difference.

From this redefinition of the Speed of Light appears the insight that information in such a Perceptual Universe could be carried by rotors in contact with one another; rotors which had an average tangential velocity of "c, " the Speed of Light.
Evenatually, this line of thoght led to the idea that Existence, as we know it, was due to Motions in a Matrix , and a paper by that name was published on Helium.com in April 2007.
It was soon realized that the basic motions would be oscillatory; and, that the matrix would have to have some flexibility, fitting the situation of a Substance at its triple point.
A Google Group was initiated in 2008, called the Oscillator/Substance Theory Group, which was quite active for three years and whose content was published by this writer in 2015, in the book, Eski's Oscillator/Substance Group, 2008-2011"
To date, Dec . 2016, there have been nine books which deal in part, or whole , with the ideas related to what is now called the Oscillators in a Substance Model. This is book number ten, which is intended to look specifically at the effect of the twin concepts of a Fundamental Substance and of Oscillators, Therein, as applied specifically to the effect on theories of atomic structure.
(Two other short books dealt with other implications, Heresy at Math.-Sci Junction dealt with some of the mathematical implications, including a new view of signed numbers, and Modern Physics Mythology dealt with some of the beliefs that the model tends to refute. This last book is available in a Spanish translation..)
While the alternative models for the atom have been mentioned in other books of this group , it was felt that the material should be reprised and elaborated as a specific topic, particularly with consideration as to how the insights of the "OiaS" model can be used in conjunction with older models for attempting to understand atomic structure and chemical reactions.
A number of concepts that are loosely defined in general usage, never defined at all, circularly defined--or otherwise needing probable clarification--receive definition from "OSM."
One immediate result of the idea of a Fundamental Substance is the idea of a single Basic Force, a very slightly varying pressure that is felt throughout. This also gives an easy definition of Gravity as the apparent attraction between aggregations arising

from the fact that the amount of the basic Material between the two aggregation would be less than the amount back of them on the same line. An even stronger argument arises for this idea of Gravity when one defines Mass as the force within the surface of an aggregation, as expressed against the Rest of Realty and felt to some extent throughout reality. Using this definition of Mass one might define the One Force as the Cumulative-Effect-of-all-Existing-Masses.

By this definition of Gravity and of Mass, it can be seen that , contrary to the belief that Gravity, or Gravity-like considerations do not apply within atoms, the force holding an atomic nucleus together, indeed the force holding all atoms together is the same as Gravity, modulated by the fact that the entities held together are spinning vortex entities (That is, by this model.)

The idea of Gravity type interactions being possible within an atom becomes important in understanding a simple model of the phenomena associated with "Cold Fusion." This will be discussed later.

The ideas of the electron and proton being spinning entities gives an explanation of Charge as a residual, net spin in a given direction. By OSM, both the electron and the proton, and their antimers, are considered as spinning, inserting vortexes which spend a portion of their time as their complementary anti-mer. This is considered to be probabably about 40%. Were there equal times spent, the particles would be identical and there would not be the s necessary for the complexity of our e , the electron is considered to have a residual net spin in one direction, possibly counter=clockwise, giving rise to what we call a Negative Charge. The proton with a residual spin which is opposite is said to be positively charged.

Three Classes oscillators, as seen by OSM

To understand the newer models of the atom which arise from OSM, one needs to know also that protons and electrons are considered as stable units of a family of oscillators which are defined by the equation, $m \times r = h/c$, with (h/c) being Planck's Constant divided by the speed of light, This Constant of Nature, which has the simulations of torque or work, may be considered the size (about $2.2l \times 10^{-37}$ g.cm.) of the basic quantum of existence. This would be the unit responsible for the name of the Quantum Revolution of the Twentieth Century.

The easier models to understand, which seem to fit the facts, are to consider the electron and proton as dual vortex oscillators, with the electron consisting of oscillators which are much farther apart from each other in frequency (and space occupied) than are the two oscillators whc make up the proton, The effect of this being that an electron can actually encase a proton both outside and inside. A perfect coordination would account for the neutron. A "looser" coordination would be a Hydrogen atom. By this view, a neutron would simply be the "lowest energy state" of the Hydrogen atom.

To understand the various models of atom, particularly as seen by the OSM model, it is well to understand a concept of three classes of oscillators which could exist in a basic substance. What we shall call a "Class One Oscillator, or a Fundamental would be a pulsating sphere with an outer and an inner limit. This Fundamental package of motions would have "harmonics," that is, other packets of motion ("Quantums of Energy") which could be added to it, one half, one third, one fourth and so on. However, addition of harmonics would change the Oscillator from a simple pulsator what we will call a Class Two Oscillator which would consist of Counter rotating halves. The most stable one of these class two oscillators would consist of the fundamental plus the second, third and sixth harmonic to be a motion packet having exactly twice the motion content (Energy) of the original pulsator. This unit would consist of two exactly equivalent, counter rotating halves. Addition of additional motion to this entity would cause it to split into two mirror-image, vortex oscillators which we shall call a Class Three Pair.

Class III Pairs of mirror-image vortices, such as the electron-positron set and the proton-antiproton set (proton/conton set) have interesting characteristics, It appears that the pairs may be interconvertible, and, at the very least, spend part of their cycle as the "opposite unit." A pair in contact with each other will form a two-center unit having existence long enough to have some chemistry as a "molecule" and then collapse to a one center form, apparently the "Original Class I Oscillator. This form in the electron set is called by this writer the Zeroton. It is to be noted that all of the above is conclusions from the basic theory . The iwo-center unit from the electron pair is known as Positronium. The Zerotron is not know n a s recognized physical uni. It is inferred by analogy and by the fact that the process of pair-production, formation of the electron positron pair under certain conditions, is also consistent with its existence.

As a basic, ubiquitous unit of existence, the Zerotron can be used to explain a number of things. In fact, it could be possibly the basic unit first created from the basic substance, or even the unit of the basic substance. The Zerotron would be expected to have the characteristics of the Class I Oscillator as described above.

It may be that the "electron-pair bond considered in molecular chemistry is a Positronium Unit which is stabilized by the presence of other entities which prevent it from reaching the exact orientation which would allow it to collapse to a single centre unit.

A major process in nature may be the collapse of two-center units to single-center units.under the pressure of the "One Force." That is, under essentially gravitational effects,. il there is not enough interference within the space between the units to prevent them from collapsing. The electron-pair bond may not only hold units together, but serve to hold them apart. It makes sense that while molecular Hydrogen, HH, with a two electron bond is stable, there seems a high probability that the Molecular cation, HH+. with a single internal electron separating two protons could collapse upon the single internal electron to form the isomeric, atomic unit, the Deuterium cation, We would see H2+ convert to $_2$H+ This will be discussed further in an explanation of the phenomena which have been studied under the names of "Cold Fusion" or LENR (Lattice Enabled Nuclear Reactions.)

About The Nucleus

It is well known that in the center ion any atom is a dense core of something. This has long been considered to be protons and neutrons held together by a mysterious "Strong Nuclear Force." By the Oscillators in a Substance Model, it appears more likely that this nucleus is made up of simpler units, none of which have free neutrons. A nucleus would be considered to be made up for the most part of Mass three units of charge one or charge two, positive. The force holding them together would be essentially the same as Gravity.

A useful view of the nucleus is to consider it as the region defined by the packing of the physically small and comparatively narrow frequency ranging protons within the envelope of the relatively huge, wide-frequency-ranging electrons. The negative charges distributed both outside and within the volume defined by the positive charges,

Blown up to dimensions we might understand, an electron would be filling a space ranging in volume from a BB shot to a football stadium while a proton occupied the space, perhaps, of a hollow tennis ball in between the BB shot and the football stadium

. By the model we are describing, the motions of either or an electron or a proton would occupy a spherical space, but at any instant, may not be a spherical unit. Were it possible to get an instantaneous snapshot of an electron or proton it could appear to be whirlwind or tornado.

Considering all of this, it appears that the descriptions of atoms by the patterns derived from line spectra, are descriptions of one particular state of the atom and energy emission or absorption, Motion pattern changes, therefore, are not shifts of one electron from one place to another; but rearrangements of the entire atom. An atom may be considered as a complex packing of spherical units. Each electron or proton moving within a spherical space. Each of these spheres in a coordinated "dance" with the rest of the atom.

The Matter/ Antimatter Problem and a Basic "Control Oscillator"

In the previous discussion we defined a Class III Oscillator set and said that the electron-positron and the proton-"conton" set would be this type of oscillators. Carefully looked at as rotating/inverting oscillators, it would seem that each member of the set would switch to its mirror image and back again. If this were true, each would balance completely and there would be no residual spin to cause charge and have a result of electromagnetism. There needs be another factor such as to cause the oscillation to be asymmetric.

This factor may well be that our universe be a result of the operation of a high frequency "Control Oscillator' at the center of our particular set of universes. If our Universe be within one lobe of an oscillator with an intrinsic spin, that spin would tend to enhance one direction of spin of basic rotors and inhibit it the other direction, so that , while each oscillator willl switch from one form to the other, the percentage of time spent ins the opposite form would be different. That is, the two oscillators would be slightly asymmetric. This Control Oscillator, as an organizer of the Basic Substance (Paleo-substance), not only gives credence to the explanation of charge, but also solves a number of other problems, the Expanding Universe, the shape and positioning of the Microwave Background, Dark Energy, Dark Matter. It, of course, leaves two fundamental questions unanswered: Why would there be a Paleo-matter? How would a Control Oscillator arise?

Related to all of this are the concepts of Matter and Antimatter. In our lobe of the Control Oscillator, we can consider all matter as composed of electrons and protons, in the corresponding "opposite" lobe, the charge effects would be reversed and we would have positrons and "contons"(Antiprotons). There have been experiments run in which "Anti-matter" was produced in our Universe in the form of "Anti-Hydrogen," which was shown to have the same characteristics as Hydrogen. There was apparently aso observed some "annihilation" of protons combining with anti-protons and electrons combining with anti-electrons to produce characteristic annular energy impulses. It was thought that Anti-matter, would annihilate with Matter to cause an explosion and, apparently, no one tried mixing Hydrogen and Anti-Hydrogen, as everyone "knew" that could cause an explosion. Much theorizing has been done as to why Anti-matter is not naturally observed in our world, numerous explanations as to why "Parity was broken." In point of fact, there is no evidence that Matter and Anti-matter would not peacefully coexist.One model of the atom which makes sense, though probably is not exactly correct, is to consider that what we observe as matter is the coordination of electrons and protons, and what we call "neutrons" is actually a cooperative, " Balancing-component-of-Anti-matter."

If it be assumed that the electronic orbital description of an atom actually describes a packing/coordination of electrons and protons, and that a corresponding description can be written for the Antimatter component which we call neutrons, this second set of written out levels turns out often to be very interesting.

For instance, this second set of units for Natural Gold, turns out to be 118 units, which would correspond to the next "Stable-Inert-Gas-Configuration."

In this writer's opinion, the whole Antimatter/Matter-Annihilation-Matter is a case of overextension from the fact that electrons and positrons, a Class-III-Oscillator-Pair, recombine to the original Class I Pulsator. In units such as the Hydrogen atom and larger units, the presence of other oscillators prevents these from reaching the exact alignment that is necessary for recombination. The electron and positron are known to form a "molecule" called Positronium and electrons are considered to form pairs in pair-bonds. It is not too illogical to consider the electron-pair bond as the same as the

Positronium unit, however, the unit is stabilized by presence of other units preventing it from reaching a configuration to allow the pair to collapse to a one center unit. That is for the potential Positronium to become a Zerotron.

The Dual Set Models of the Atom, Outer/Inner or Matter/Antimatter Sets

A very logical, and useful alternative to the idea of there being neutrons in nuclei is what may be called "Dual Set Models of the Atom. " The idea is that what has always been observed is one set of electrons correlated to a set of protons, and, unobserved, and seen as neutral units, is a second set which has been erroneously considered as consisting of neutrons. This second, correlated set of electrons and protons, which could also be considered as anti-electrons and anti-protons ("Positrons and Contons") operates as balance; but, because of lower motion content may need more units. Both sets may be considered to have spacing;"orbitals" described by the designations given to the outer set by reference to line spectra.

As the element, Nickel, is a very important industrial catalyst, and Ni62 is considered one of the two most stable of all isotopes, the natural isotopes of Nickel will be discussed here as well as the natural isotopes of Copper, Cu 63 and Cu65.

Element 28, Nickel, has naturally occurring isotopes of mass 58, 60, 61, 62 and 64, The supposed electronic structure, called in this modelling, the Outer Set, or Matter Set, $1S2: 2S2, 2P6: 3S2, 3P6; 4S2, 3D8$. This set of 10 outermost electrons makes a very interesting situation. It is very close to being a situation of $4S2, 3D5, 4P3$ which would have two half-filled sub-shells, 3D and 4P, in addition to the filled 4S sub-shell. Ionization of the two 4S electrons would still leave the two half-filled sub-shells.

If one is not familiar with the Shell-subshell notation of this type of model, this may be simply science jargon without meaning, Here is an attempt to clarify. Every element has a set of light frequencies which are characteristic of that element. Those frequencies are divided into sets which are called line spectra. A set of one line called the S, for Sharp, line which is said to be due to the presence of a set of 2 "electron orbitals," places where one set of electrons could dwell; a set of three lines, called the P (Principal) lines, which are considered to be another set of Orbitals which can contain up to six electrons, a set of 5 lines, called, D for Diffuse, which is considered to indicate a set of 5 Orbitals, which can contain up to 10 electrons, and a set of 7 lines called the F, for Fine lines, which supposedly result from 7 orbitals holding a possible 14 electrons.

According to this model, electrons arrange themselves in atoms in shells and subshells, the first shell, called the Level One, or K, shell, supposedly has but one sub-shell, an S level containing, at the most, two electrons. When this shell is filled, at two electrons, there arises what is called a noble gas, without much chemistry, Helium. The next Shell, has two levels, S, and P and when these, and the first level are filled, there arises, at Element 10, another noble gas, Neon. At the third shell there arises a third subshell, which does not start filling until after the 4th S sub-shell. In other words, the pattern gets a bit more confusing. An attempt will be made later to clarify this with a chart. The thing to remember is that, by this model, which seems to make reasonably accurate predictions as to elemental and ionic stabilities-- whatever relationship it may have to reality-- filled and half filled shells and subshells are more stable than arrangements whc do not fit as one or the other. The NI++ cation would be, at first glance, 1S2: 2s2, 2P6: 3S2, 3P6: 4S0. 3 D8.   However, the last two levels, by this notation. are an empty 4S and a 3D which is three more electrons than a half subshell of 5 electrons. However, these three extra electrons would correspond to a possible half filled 4P subshell.dd, giving an alternative configuration with nearly the same energy content of 4S0, 4P3, 3D5. For eight unpaired electrons. The Nickel di-cation could have any set in between these two extremes. Unionized Nickel could also have any set between the extremeriences of 3 D8 and 4P3, 3D5 for the arrangement of its   sets of electrons. This probably is a factor in the fact that Nickel metal is a catalyst for a number of reactions.

Let us look at the internal electron situation by this model with the situation of the different isotopes (weight variations) of NIckel, Nickel 58 would have t0 electrons in its interior 4D shell, giving a final grouping of 4S2, 3D10, both subshells filled, a stable situation according to the idea that filled and half filled subshells tend to be stable situations. Ni 49 is too unstable to exist in nature, However, Ni60, with the supposed last two inner levels being 4S2, 4P2, with two sets paired electrons, or , very possibly 4S1, 4P3, with 4 unpaired electrons, in two half filled subshells, is stable. By the same figuring, Nii 61, with a filled 4S and a half-filled 4P should be stable, and it is. NI 62, with an additional 4S electron producing a filled subshell, is very stable, In fact.\t is, along with an Iron isotope, considered as one of the most "stable of all isotopes.   Ni63 is not stable with respect to Cu63 and does not exist naturally. Ni64 is the last of the naturally occurring forms with the internal notation for the last internal levels of 4S1, 4P6, a half-filled and a filled sub-shells.More discussion of these various nickel ions and their significance with respect to the problems of the Cold-Fusion/Lattice-Assisted-Nuclear-Reactions Area will be done later; when the insights of this model, and of another model called the "Four Factor Model " of atoms will be combined to try to explain what may be going on in that particular controversial area of the chemistry of Hydrogen.

Back to Hydrogen, the First "Element"

It is very possible that the last few paragraphs dealing with NIckel, have been a bit difficult to understand. The chemistry of Hydrogen, the first element, has been mentioned. Let us go back and examine Hydrogen, the first Element.

Most of Chemistry is organized around the concept of Elements, units which have a common chemistry although they may not have the same size/weight. The first Element is known as Hydrogen, literally, the "Creator of Water." It is made up of three units having masses of one mass unit, two mass units and three mass units, all of which react with other Elements in such a way as to apparently, in most cases, share one electron with some other element. (These are the natural forms, there are reported a number of other forms from particle accelerator work. We shall ignore them.)

In forming water, H2O, two Hydrogen atom share one electron each with an Oxygen atom to form the compound, Hydrogen Oxide, Water. Water usually is made up of Oxygen 16, and Hydrogen 1, for a weight of 18, but that is not always the case, Oxygen may have other weights, There are a number of units which can be called "Water." For most purposes this is ignored, as the other units other than mass 18 are usually in such small proportions as to be negligible in importance. However, sometimes, this differentiation can be critical. More on this, later.

Hydrogen I, $_1$H. Protium, is considered to be formed of one electron, and one proton. Hydrogen 2, $_2$H, Deuterium, Is usually considered to be formed from one electron, one proton and one neutron.

(The writer prefers a model containing two protons and two electrons; or another alternative containing a proton and antiproton and an electron and an antielectron.)

Hydrogen 3, $_3$H, Tritium, is generally considered as made up of one electron, one proton and two neutrons. Again this writer disputes the presence of neutrons in this unit and suggests that better models can be formed from a two set model. Tritium is radioactive, emitting an electron (Beta particle) to convert to a Helium 3 mono-cation. This is the first case of a mono-cation which could, by rearrangement of its internal electrons, convert to either of two units.

If Tritium were simply to ionize, the mono-cation of Tritium would be said, by the two level model as used in the discussion of Nickel done above, to have an outer configuration of 1S zero, and an internal configuration of 1S2; Empty outer shell, paired inner.shell. The energetic ionization gives the more stable cation configuration, outer, 1S1, inner also 1SI, half filled outer and inner shells. By this model, the two apparently different cations would be interconvertible by the shift of one electron from an outer to an inner set. This inter convertibility of the Tritium and Helium Three cations, leads to a model which we will discuss later called the Four Factor Model of the atom, which the writer considers the most realistic model of atomic structure. That model considers that virtually all isotopes may be defined as formed from atoms of Deuterium, Tritium and Helium Three.

More specifically, the nuclei can be formed from the He3++/T++ common di-cation, the He3+/t+ common cation and the D+ cation. In some cases, the Proton, H+, is also involved, and in a very few, very rare and extremely short-lived units, neutrons are apparently actually involved, This is in isotopes with half-lives far shorter than that of the neutron itself. These last, peculiar entities are probably unknown in nature.

The point to remember here is that the three naturally occurring isotopes of "Hydrogen:" may be considered as the building blocks of all of the rest of matter. While the idea that there be outer and inner sets of protons and electrons makes a model which can be used to explain much chemistry and nuclear chemistry, the four factor model seems to be closer to reality.

In practice, the two seem to reach the same conclusions much of the time, with one sometimes pointing out an idea which is not immediately apparent from the other model. The inner/outer sets idea tends to emphasize that electrons may have a great deal of freedom of motion within atoms while the four factor model would tend to imply the locking of some electrons into rather fixed positions. Reality is probably somewhere in between.

The naturally occurring isotopes of Lithium, Element 3, and of Beryllium, Element 4. offer interesting chances to compare the two models and their interactions. Lithium has two naturally occurring isotopes, Li 6 and Li 7, Beryllium has only Be9 as a stable, naturally occurring isotope. Lithium 6 has an outer electron coding of ! S 2; 2S1, and an inner coding which would be the same. A situation that looks nicely stable. Beryllium 6, would have outer shell sets of 1S2; 2S2 and an inner shell set of 1S2. This is an unbalanced situation with respect to the Lithium 6 situation. Similarly Be7 would be 1S2, 2S2 outer and 1S2; 2S1, inner. Which is unstable with respect to the Lithium situation, which is the reverse. There appears to be a rule in nature, using this model, that the inner set of a stable unit will have either the same or more electrons than the outer unit. (This corresponds to the proton/neutron model of "neutrons" being equal to or more than, the number of protons in the nucleus in naturally occurring elements. (Except, of course, in the case of Hydrogen 1 where the "neutron" would be a special cale of a totally compressed H-I and would, if looked at that way, still follow the rule.

The set, LI 8, Be 8, B 8 (mass number 8 of Elements 3, 4 and 5) are interesting. Li8 would have outer, 1S2; 2S1, which defines it as Lithium, Element 3, and an inner of 1S2; 2S2. 2P1. This is unbalanced with respect to the Beryllium 8 set of 1S2, 2S2, both inner and outer, Li8 converts to Be-8. On the other side, B-8 is similarly unbalanced to the outer side with 5 outer to 3 inner electrons by this modelling. It, too, converts to Be8.

As Be8 is stable with respect to both LI8 and B8, we would expect it to be very stable. Yet, in practice, it decomposes--emitting an Alpha particle--and, eventually, becomes two units of He4. Element two, mass four. The emission of the Alpha particle, mass four, charge two, positive, gives a clue as to the probability that the instability is not of neutral Beryllium atom but of the quite easily formed Beryllium dication, Be++ . This is a unit which would have the external configuration of 1S2, 2S -0 (zero, shell, empty) and an internal 1S2, 2S2, two filled shells, If these total six electrons happen to momentarily distribute as 4 in one half of the unit, two in the other, these disparate halves can, and apparently do, split into two separate, individually stable, very dissimilar Units: A vibrating unit which is essentially He4; and a rotating di-cation, the Alpha Particle, which spins out and away, eventually picking up two electrons to become another unit of He4.

 The Alpha particle is sometimes called the He4 nucleus. This writer is of the opinion that the Alpha is probably better considered as isomeric to the He4 nucleus, but not identical. The he4 nucleus is possibly a vibrating tetrahedron whereas the Alpha is almost surely a planar rotor. It is highly probable that the Alpha Decay of higher elements is a characteristic of the di-cations of those elements and not of the neutral atom nor of a monocation. It may well be that the one naturally-occurring Bismuth isotope is essentially stable as there seems little tendency for this Bismuth isotope to form free cations, of any sort, certainly not a di-cation. In addition, the internal set ("Neutrons") is the same as that of maturally occurring Gold, the same as the expected outer set for the "Next Noble Gas."

The Shell/Subshell "Stepwise-Filling" Chart.

For determining probable shell and subshell configuration for electron orbitals an easily memorized and reproducible chart is useful . his chart is shown below with an explanation .

Shell Number	Sub-shell Designation			
	S	P	D	F
1	1,1			
2	1,1	3,3		
3	1,1	3,3	5,5	
4	1,1	3,3	5,5	7,7
5	1,1	3,3	5,5	7,7
6	1,1	33	5,5	7,7
7	1,1	3,3	5,5	7,7

Each shell is said to have the same number of possible sub-shells as the number of the shell. That is Shell One has one sub-shell, Shell Two has two sub-shells and so on. These sub-shells have a capacity of twice the prime numbers associated with them. Two for the first shell and sub-shell, six for the second sub-shell, ten for the third and fourteen for the fourth. These odd numbers are all primes, the fifth shell would be expected to have a sub-shell with a capacity of 18 units, twice the next odd number of 9. However. nine is not a prime.

At this point the situation gets complex, apparently because nine is not a prime number. Luckily, we run out of known Elements before that sub-shell be needed. Using the above chart to write shell and subshell designations for isotopes of Nickel, Element 28, as the element-defining outer set, we obtain 1S2; 2S2, 2P6; 3S2. 3P6; 4S2, 3D8.

A careful consideration of the above notation and the chart will show how the chart works as a mnemonic device for each grouping of electrons starting at the "upper North East" and continuing down to the "lower South West."

It is known that half filled subshells are also situations of stability so one might suspect that instead of the last set being 3D8, we might have an almost equally probable 3D5, 4P3 situation of two half filled shells and eight unpaired electrons. This suggests a possible connection to the well known fact that Nickel is an excellent catalyst for reactions involving Hydrogen, very possibly sometimes involving the splitting of Hydrogen Molecules into Hydrogen atoms.

The internal sets of Nickel are interesting. If we look at the situation with Nichel of mass 62, for its 34 internal sets we find that our last sets be 3D10. a filled shell, and subshell, and , probably, 4P3 and 5S1, two half filled subshells. A very stable situation for an interior. The situation for Ni64 is also interesting as its internal set would have a completely filled 4P subshell in the interior. A nucleus that would have a possibility of temporary addition of two units at 5S. Internally.

These ideas will be explored further when we look at the probable Nickel and molecular Hydrogen interactions that seem to give a reasonable explanation for the "Mystery of Cold Fusion" discovered a quarter of a century ago by Fleischmann and Pons when they had a Palladium electrode melt down while electrolyzing "Heavy Water" and observing the production of Helium 4 and excess heat. Although the original work was done with Palladium and D2, it is simpler to try to explain the Cold Fusion situation with the results of later work with Nickel.

As another example of use of the above step chart, we can write the outer set for Palladium, which is considered to be a member of the same set of elements as is NIckel. This is the set known as the Catalytic Elements, Nickel, Palladium and Platinum. By our step chart, the Outer Set, which defines the Element, Palladium, is composed of 46 units, with the folloeing, expected configuration, 1S2; 2S2, 2P6; 3S2, 3P6: 4S2, 3D10, 4P6; 5S 2, 4D8, However the set of 5S2, 4D8 is so close in motion dontent to an empty 4S and a full 3 D that the latter is usually considered to be the ;outer cconfiguration.It may also noted that the set 5S2, 3D5, and 5P3, would also be very close in motion content and may in volved in catalytic and nuclear transformation processes. Comparing the above description to the chart given above should help in understanding the use of this "Stepwise, level-filling chart." This chart can be used to used to describe possible electron orbitals (packing patterns by the Oscillator Substance Model) for any element or postulated element, to estimate probable ions; and , also. possible nuclear transformations by doing the sets for both the "Outer set" and "Inner Set," as was discussed earlier in the case of certain isotopes of Elements 2, 3, and 4. (Lithium, Beryllium and Boron.)

Introduction to the Four Factor Atom Model

Another model which has arisen from the OSM work, but which could, perhaps, have been discovered independent of the Oscillator Substance Model, is the Four Factor Model of the Atom where atoms are considered to be made up of the forms of Hydrogen , and Helium 3, which is isometric to H#\3. The observation was made that virtually all reported isotopes, except for a few very-short-lived, "Halo" atoms, could be considered to be made up from units of Helium Three, Tritium, Deuterium and /or Protium (Hydrogen One.) Probably it is better to state the equivalent idea that nuclei, instead of being made up of protons and neutrons, be actually made up of clusters of ions which may be called by the following symbols, 3++ (The Helium 3/Tritium Common di-cation); 3+, (the Helium 3/Tritium Common mono-cation), 2+ (Deuterium

Cation.) and 1+, the proton, Hydrogen 1 Caton.) These are charted in that order, so the nucleus of, Lithium 6, for instance, could be charted as "Isotope 1,1,0,0." One 3++ and one 3+. with the other two units indicated as absent. Helium 4 Nucleus would chart as 0.0.2,0 and a proton alone, would be 0,0,0,I. by this notation, a huge majority of reported isotopes can be positively identified by a combination of 4 numbers.

(It is an interesting coincidence that Gold, sometimes associated with the Sun , has its naturally occurring isotope chart as 13, 52, 1, 0. This seems a recipe for a solar calendar of 13 months of 52 weeks with one day over. There appears even a hint that the confinement to four symbols doesn't quite tell the whole story.)

It turns out that the number of Deuteron (2+) units may be zero, one or two. With isotopes having one Deuteron apparently somewhat more likely to be stable than neighboring isotopes of zero or two 2+ units. A situation with three Deuteron units apparently has no stability.

It is as if a single Deuteron can operate almost as if it were a "balance pole" on the tight stability line of the isotope. Other interesting patterns also arise. It appears that one #++ unit is necessary for an isotope of an element (above Element 2, Helium) to exist, and also there must be at least one less of the 3+ units for an isotope to be stable . That is to guess at what is likely to be the configuration of highest isotope of some element , one would divide the atomic number of the element between units of 3++ and 3+ such as there would be one less of the 3+ units. If we wanted to estimate a lowest mass value for Element 23, choosing an element at random, we might assign all of the charges to mass Three units such that there would be one less of the 3+ units, and note that 2 x 8 + 1 x 7 =23, hence we would decide that a lowest mass number expected for Element 32 would be at tmass 45 , Even simpler, however, is to go back to the old proton-neutron ideas and note that the number of "neutrons" in never less than one less than the number of protons, which, of course,is the atomic number, so the lowest weight isotope expected for element 23 is 2 x 23 minus one, or 45. The old proton-neutron bookkeeping is useful. Similarly for Gold, Element 79, the lightest isotope would be expected to be mass 157. At the other end of the scale, to estimate the maximum mass number for an isotope of an element we could go to the idea of at least one 3++ being necessary and all of the other positive charges that make up the atomic number of the element being associated with 3+ units, that is , all of the mass being in units of three. This would lead to the idea that this mass would be three times the atomic number minus one. That is for our Element 23, the heaviest expected isotope would get the isotope of mass 66. We might guess that the most stable isotope would be somewhere close to the average of these two figures, 45 and 66 which is 55.5. We might even guess that there could be two stable isomers at 55 and 57, perhaps.

Looking up element 23 Vanadium, we find that the estimate for the heaviest known isotope at 66 was absolutely correct. At the other end of the scale, very light, very short lived isotopes have been reported beginning at Isotope 40 rather than 45 and the most stable situation turns out to be at a single isotope , Isotope 51. It would appear that our presumption of the heaviest isotope being composed of units of three containing one 3++ unit, is accurate, at least in this case. The no-less -than-one-less rule needs some reexamination and the estimation of stability from the arithmetic average of lightest and heaviest appears only a rough approximation.

(The above few paragraphs illustrate the procedure of this kind of research. We use data to develop a model. Check for predictions of the model, see whether those predictions seem to be accurate and look for modification if the are in error. The prediction of the heaviest known isotope for Element 23, Vanadium, an element chosen at random, seems evidence that the assumption made for the heaviest isotope be correct. However, we cannot now assume that the assumption will hold for all elements without checking each and every one. At the other end, our assumption that there will not be found isotopes lower than a situation of one less 3+ unit than there are 3++ units appears to be in error. Another assumption we could "try on for size" is that the situation at the low molecular weight ernd for an element is that it may be essentially the reverse as at the upper end. That is that unstable units may exist as long as there be present one 3+. unit.
'

Checking out the four factor coding for Vanadium of mass 40 we obtain two possible codings. 10, 1, 0. 1. Or, as an alternative structure, 10, 2,2,0. This situation of two possible codings indicates that there may be two isotopes of the same mass which can undergo "internal conversion." Internal conversion is a situation which is well documented for many other isotope pairs having this coding possibility. .\ Here we have a case of three 3+ unit supported by a 1 unit or of two 3+ units supported by two 2+ units.

Another way the four factor coding could be used could be in looking for the possible stable isomers for a particular total weight. We can start with our B40 and write alternate codings for mass 40, starting with the B40 as 10, 2,2,0, we can go to 9,3, 2 0; and so on eventually arriving at a coding of 6,6, 2,0. Which has equal numbers of 3+ and 3++ units, and could very likely be stable. This would be an isotope of Element 20 .Looking up element 20, Calcium we find that , while Ca40 is not totally stable, it does have a very long half-life. Long enough for Ca 40 to occur in nature. It turns out there are, also. two completely stable isotopes at mass 40. These are the coding of 5.7.2. 0, an isotope of Potassium, Element 19; and, 4. 8. 2. 0, an isotope of Argon, Element 18.

This coding suggests possible families other than the families which we know as elements, that is, families of the same outer coding by our two set codings. An obvious one is the type of family we have explored by following a possible set of changes for mass 40 from a "start" as light isotope of Vanadium to isotopes of Calcium , Potassium and Argon. This was part of the "Mass 40" family. Following Mass 40 to a logical end, we could expect it to appear as one of the the heaviest isotopes of Element 15 with at a coding of 1, 11, 2,0. It turns out that while this isotope of Phosphorous is reported there are units up to mass 46 reported. An isotope of Phosphorus of mass 46 would be a strange unit coding as having Zero

3++; 15 of the 3.+ units Zero 2+, and an apparent mass unit without charge, this could be a "halo" isotope which actually does contain a neutron. Another possibility is that there is present another rare nuclear unit, in this case, a 4 + unit.

If we abandon the idea that a 3++ unit be necessary, we can check to see if a unit of mass 40 coded as 0, 12, 2, 0 be reported for Element 14. A check of the literature shows this to be the case, and for, Silicon, Element 14, isotopes are reported to weight of 44. This, again is a situation which does not fit with the four-factor coding that holds for most isotopes. A possibility, here again,' is that there may be at the very high end a situation of a 4+ unit having existence. It appears that our 4 factor coding works well for everything except the almost ephemeral units to be found at the lighter and heavier end any given element set. As can be seen from the exploring done while writing this, our 4 factor model is valid for most of the isotopes known, but fails at the extremities. The "Two Set" models--discussed earlier-- may be used to rationalize these extreme cases.

We can look at these other models that we discussed where we use the electron orbitals type of levels, but write structures for both inner and outer structures. We found that our four factor coding failed at the extremities to account for a few isotopes as, for instance, isotopes 41 to 44 of the Element Phosphorous, Element 14. The two level model, however can account for these additional, very short lived isotopes as having inner arrangements that are the same as outer arrangements of other elements. Isotope 44 of Element 15 would have an internal set, by the two set model, of 29 units, which is the same structure as the external structure written for Element 29, Copper, a unit which would be expected to be quite stable. It appears that the four factor coding offers very useful insights for isotopes other than at the extremes.

A Short Note. Neutron Stars as Over-sized Atoms

While It seems logical to model "neutron stars" as over-sized atoms, the fact that unusual units appear at the heavy end of element chains, implies that processes at the extreme levels of an "atom" having many millions of units would be somewhat beyond the rules that seem to govern the more stable units. Considering the "One Force" concept of the Oscillators in a Substance Model, a " neutron star" could be made up of a nucleus containing many nuclei far beyond the four that we are considering. That is, the nucleus of an atom, and of a "neutron star" could have more resemblance to the structure of a rocky planet than we might suspect.

The postulate in OSM of an ubiquitous Zerotron, distortable to neutrons by shock wave processes. suggests that a neutron star might well be continuously generating neutrons. Those neutrons would decay to electrons and protons, allowing the star to continuously regenerate itself. The star would not consist only of neutrons, however.

Our Models and the "Problem" of Cold Fusion

The reader's indulgence is asked, Rather than "starting from scratch" for this section, this writer is taking the privilege of doing some "self-plagiarization" by reprinting here, with some edition and addition, a previous article on cold-fusion published in another of his books. (<u>Off the Wall, Vol. 1</u>.)

Comment on Cold Fusion from the View of the Oscillators-in-a-Substance Model

Summary: The Oscillators in a Substance Model of Existence (OiaS, O/S, "Waas,"OSM) which considers that orbital levels in an atom actually correspond to packing patterns for spheres, offers an interesting possible explanation for the transmutations, heat generation and Helium 4 generation observed in "Cold Fusion " experiments. This explanation indicates that what is being seen may be previously-unnoticed chemistry of the units considered as "Hydrogen" with cationic species which have what may be called "Open Spaces" within their atomic structures. This would appear to apply to cations of the transition series elements. The situation with isotopes of Nickel is discussed as an example.

In the Oscillators in a Substance Model of Existence, (henceforth, herein, abbreviated to the sound of "OiaS" as "Waas") there are essentially two basic tenets, "Existence is considered as being within a Substance;" and, "Within that substance are oscillations of a family having a basic work function of about 2.2 x 10^{-37} g.-cm. That is, oscillators belonging to a family of oscillators defined by the equation, m x r = h/c. (1)

A result is that electrons and protons are considered as vortex oscillators occupying spherical spaces and packing as spheres. The vastly different sizes of the two oscillators result in a model of an atom as having a dual structure with an internal, tiny sphere (the nucleus) occupied by the proton units; and, a much larger outer and even tinier inner sphere--within the nucleus --which are defined by the motions of the electrons. The electrons have a much wider range of motion allowing electrons to have packing such that parts of the electron vortex be within the nucleus.

Both the electron and the proton may be considered as dual oscillators of widely different frequencies (wavelengths, sizes, masses) .

This becomes important in modelling the internal transforms that appear to account for "Cold Fusion."

While the Cold Fusion Process has been studied to the greatest extent in the original context of deuterated water electrolyzed with Palladium electrodes, the situation with the Isotopes of Nickel will be considered here.

There seems to be reasonable support for reports that all Nickel isotopes, except Ni 61, can "participate in Helium 4 production and excess energy production under electrolysis conditions with simple water. It appears, also, that the report that the "ash" (Nickel recovered from these experiments) is only Ni 62-- seems likely to be valid. Assuming these results to be correct and that the patterns possibly be general, the following is proposed by extending ideas from molecular chemistry to the nuclear level.

Nickel, as a transition metal, will have within itself "open energy levels." That is--in Waas reading-- open space into which electrons and protons can be fitted, Cationic forms will be even more vulnerable.

In Molecular Chemistry any cation will have the potential to act as a Lewis Acid--an electron pair acceptor,

H_2, molecular Hydrogen, or DD, Molecular Deuterium, has the potential to be a Lewis Base--an "electron-pair donor. Both Molecular Hydrogen and Molecular Deuterium--when converted to their "single-point--centered, isomeric forms," (respectively, the D atom as an isomer of the H_2 molecule and He4 as an isomer of the DD molecule) are potential internal units to be added for production of higher weight atoms. By the models worked out from the Waas reasoning, the Deuterium cation, D+ ($_2H^+$) is one of the possible components of atomic nuclei.

A derivative of the Waas model is the observation that any naturally occurring isotope which is larger than the "3-pair" of Tritium, H3, and its isomer, He3, may be considered as being formed of combinations of D, T or He3. This also applies to virtually all of the other known isotopes . A few rare isotopes need to have also $_1H$ as a possible component, and a very few, extraordinarily-short-lived isotopes seem to have neutrons included within them.

Closer examination of the above indicates that it may be more accurate to consider that what is found are not the neutral units but corresponding cations in the nuclei. The nuclei of most atoms can be "coded" as units of "3++, 3+ and 2+ corresponding to a He3 or T di-cation, a He3 or T mono-cation and a Deuterium mono-cation. An example is "Natural Gold," Atomic No. 79, Atomic Wt. 197, which codes as "`13, 52,1" (Thirteen of 3^{++}, 52 of 3^+, and one of 2^+) (I)

The "3++" units may be considered to have one "embedded" electron, each, of the "3 +" cations have two embedded-electrons and the 2+ to have one embedded-electron. The interesting result of this analysis is that the total of the embedded electron--not only for Gold; but, also, for any isotope, corresponds to the number of "neutrons,: in conventional modelling.

, When this number of embedded-electrons, is charted for Gold, by the charting that is usually used for "electrons in levels." it is found that the 118 "embedded electrons ," correspond exactly to what would be expected for the outer configuration of the next "inert gas."

With the waas Model, there arise several alternative ways of considering atoms and atomic structure depending on the viewpoint and the need of the time. It is not to be considered that any one of these views is "absolute truth," but each can be useful.

the "proton-neutron. nuclear atom" is so pervasive in the literature of the last three quarters of a century , that anyone who has studied physical science knows it. It is useful for bookkeeping Purposes but should not be taken seriously as saying that neutrons, as such, exist in most atoms.

[Note: If there be no neutrons, as such, in the nuclei of atoms, then how are they apparently being released in well known fission reactions? Wass theory suggests two possible explanations. One is the shock wave distortion to neutrons of a logical--but unrecognized--neutral unit, the "Zerotron," which may be pervasive to the "void."(1) Also, the commonest unit of nuclei, by this model, is the 3+ unit, which may be quite easily fissioned into 2^+, the deuteron. and $_1n^0$. the neutron.]

The model of two inter-penetrating droplets with their individual surfaces is a view to keep in mind.

The " 1s, 2s, 2p, etc," electron-structure-charting is still very useful, especially if one considers that it may be used to represent either one of two possible coordinated sets, The first being a set of electrons and protons. and the second being another coordinated set of ""anti-electrons" and "antiprotons."
This "Two-Sets View" may be taken as saying that every atom is made up of cooperating packages of matter and anti-matter. The "anti-matter package" being previously considered as made up of neutrons.

As no one is used to thinking of matter and anti-matter co-operating rather than annihilating, there is probably a better "comfort zone" in calling the second set of levels, "inner electron levels," rather than either neutron levels or anti-matter levels. In truth, it is probably closer to think of both as "packing patterns." and to consider that "energy level transfers" probably represent a shift in position of every part of the atom rather than relocation.of but one electron.

How does this fit into "Cold Fusion and Nickel," which is what this paper is about? The natural isotopes of NI, are mass numbers 58, 60, 61, 62 and 64. with NI62 considered by many as being the most stable of all known isotopes. The only isotope that may remain under some conditions of a reaction involving Nickel and Hydrogen appears to be this Ni 62.

We can postulate the following series of reactions to account for a part of this. as follows: $Ni\ 58^{++} + H_2 \rightarrow Ni60^{++}$; $Ni60^{++} + H_2 \rightarrow NI62^{++}$.

The postulate is that the tiny $_1H_2$ will coordinate, not just at the surface of the NIckel cation, but will actually penetrate the other surface, dropping an electron into the "electron packing pool" and producing a H_2^+ , this cation, under the pressure of the "One Force of Nature,"(1) as expressed within the atom--continues into the nuclear region with the two protons collapsing upon the remaining electron (the high-frequency, high-mass part, thereof, if one considers the electron as a dual oscillator,) The new 2$^+$ cation takes up residence within the nucleus and the unit stabilizes to a heavier isotope of Nickel. The above ideas work for 58 and 60 to get to 62. We still have mysteries with isotopes 61 and 64.

What happens with 61? How does 64 backtrack to 62? Where does the He4 fit in?

Another idea noted in the "Waas" model is that there will be a common cation between any two isotopes of the same mass, which are interconvertible by the movement internally of an electron. The first case would be the Tritium/He3 set, and all subsequent ones would be referred to this. Ni 61 plus H_2 produces an unit which would have common cation to Cu 63, which is naturally occurring, long-term stable. Nl 63 is not.

Unfortunately, no one has as yet checked for the presence of Copper isotopes which would indicate this as a feasible postulate. For the moment, however, let us assume that this transform accounts for the fate of Nl 61.

what about the Nl64 situation? Nl 64^{++} plus H2 would produce a cation which would not have an easy transform to a stable nucleus, neither Ni66 nor Cu66 is naturally occurring. Going to the ",3,3,2" coding, we find that that there could be a possible. momentary set of 3 units of 2^+. This unstable structure could convert to a unit, 4^{++}, the Alpha Particle, and leave one 2^+ ($_2H^+$) behind.

(By the coding, naturally-occurring, long-term stable, units may have zero, one, or two "2^+" units, but never more than two.)(l)

The appearance of Ni 62 and Helium 4 indicates a possible cycle with $Nl66^{++}$ formed momentarily, but immediately fissioning, releasing an Alpha unit (6/8 of a He4 unit.) The detection of a few Alpha particles supports this.

Addition of molecular Hydrogen to Ni62 would form Ni64 which could go to Ni 66 as noted above. Subsequently, an Alpha particle going to He4 could abstract electrons, creating more Nickel cations. One can see a repeated cyclic reaction with the effect of doing a fusion of two H2 molecules to an He4 atom. This would be a cyclic process of two successive fusions followed by a fission, definitely a situation which would produce excess heat. Atomic Deuterium could "shortcut" the process and molecular Deuterium, even more so.

If the above ideas of the reaction of molecular Hydrogen with cations, primarily the di-cations of transition elements be valid, then there is a definite possibility that intra-elemental transmutation processes may go on in electroplating processes and other electrolyses; but, have never been noticed. We have hinted at one possible example.
In the above discussion, it has been noted that the probability is that Ni 61 disappears to form Cu63. While Cu63 is naturally occurring and "stable," analysis by the models that occur in Waas suggests that the Cu 63 could transform to Cu65 which may be resistant to further change. (Its "inner structure" corresponds to an "inert gas structure.") This suggests, however. that electroplated Copper might be predominantly--or totally-- Cu65. There seems a slight possibility that $Cu65^{++}$ could catalyze Deuterium formation.
The idea that the "Cold Fusion" transmutation processes may be a quite general reaction of Hydrogen forms with certain cations indicates that

there may be many unnoticed uses beyond Helium formation and heat production.

One suggestion seems to have been made in a patent application that this type of process could possibly be used to "cool down" nuclear waste by transforming radioisotopes to less active or stable forms.

The above paper does not go into the possibility that there may well be very special conditions for the interaction of the very weak Lewis Base represented by Molecular Hydrogen forms and the much stronger Lewis Acids represented by the di-cations. Dr Edmund Storms is probably correct in feeling that the size of cracks on the catalytic surfaces whether Palladium or Nickel may be somewhat critical. It may simply be that those cracks must be large enough to allow HH, DD, or HD to accumulate; but, small enough to prevent the entry of stronger Lewis Bases, e. g. water, to enter and block out the Cation/Molecular Hydrogen interaction. Also there is the problem of the necessity of there being active di-cationic sites at the electrode, which would be expected to be negatively charged.

Another factor is that the presence of significant HOH in DOD seems to "poison" the reaction when using Palladium and "Heavy Water;" and, significant

DOD poisons the reaction with simple water and Nickel. What is probably the problem is that any DH, mass three, Molecular Hydrogen, entering into the stage which would produce Alpha particles would add three units to the nucleus rather than two or four. That is, if the HD+ cation collapsed to a 3+ unit, it would probably convert Ni62 to Cu65, destroying an active site. Presumably, the same type of situation holds with Palladium isotopes. In other words, if you want to use Cold Fusion to produce Heat and Helium, you need to work with as pure HOH, or DOD, as possible to avoid problems with the mass three Molecular Hydrogen, HD.

In recapitulation/summarisation of the above article:

The phenomena associated with "Cold Fusion" seem to be understandable if one looks closely at the implications of the Ideas of the Oscillators in a Substance Model. One key factor is the One Force which will cause essentially Gravitational forces to exist even between such small units as electrons and protons and can be considered to interact at such small distances as within atoms. That is, the idea of a molecular Hydrogen unit being absorbed into a larger atom and being attracted to the nuclear region much as an asteroid might be attracted to a planet makes sense. That this neutral unit, entering into a Dication might become a mono-cation (charge/spin balancing) makes sense and the compression of this resulting two center unit to a single center unit, by collapse also makes sense. That is the idea of the sequence HH-->HH+-->D+, or a similar sequence starting with HD or DD, makes sense, if we assume that simple mechanics actually holds quite well no matter what the matter unit is or the conditions.

That the converted unit on collision with the nucleus may suffer perhaps three fates also makes sense. The unit may be easily incorporated into the nucleus causing a fairly simple transmutation. It may undergo an interaction with the nucleus with a subsequent emission of another unit; or, it may simply "bounce" on collision with an especially symmetric and stable nucleus.

In the discussions with Nickel the transform of two Hydrogen molecules to Helium is considered to take place from Ni62 by way of a "peaceful incorporation" of one D+ unit, and subsequently more turbulent incorporation a second unit, combining with one of two already present D+ units to create a "D+D+" Unit, an Alpha particle, which spins off.

If instead of starting with the Hydrogen One dimer, we started with Hydrogen Two dimer, so that the produced "Cation-to-hit-the-nucleus" be the 4+

cation, one can easily see this unit, in bouncing off the Ni62 nucleus, dropping an electron and leaving as the Alpha Particle, "4++." It may well be that this is the situation wlth DD and Cations of certain Palladium isotopes. Molecular Hydrogen forms being able to be incorporated into cations would depend on those cations not being blocked by other chemical units, which brings us to the probable reason that this phenomenon has only appeared in certain cases of electrolysis work where it is probable that surface crazing creates areas in which Hydrogen Molecules can reach "unobstructed" cationic sites. In other words, the surface condition is not the mechanism, but the physical condition which allows the mechanism to take place.

A part of this mechanism, the collapse of a molecular monocation to the atomic isomer, could happen, whenever, and wherever that particular unit would have a chance to form or any time when the two electron bond separating two units be reduced to a single electron, allowing the two portions to collapse upon the central electron, If O2+ were ever formed, it would be expected to collapse to S+, etc. This formation of mono cations of diatomic molecules is a very rare possibility in nature but the idea of collapse taking place in some unit might not require the actual formation of this type of unit. An

Italian researcher has mentioned the observation of Sulfur in anode cells during electrolysis with Copper electrodes. He did not say what species was noted. However the fact that it was with Copper is very interesting. Copper, in the Cuprous form is a dimeric form, Cu_2^{++}. with two unpaired electrons. Molecular Oxygen, O_2, is also a dimeric unit with two unpaired electrons. It seems very likely that there could be formed an unstable unit, a four membered ring, (Copper Peroxide) Cu_2O_2, which by shift of one electron around the ring, and collapse together of the two Oxygen nuclei, could "disproportionate" to free Copper unit and Cupric Sulfide, CuS.

(It may be possible that adding Hydrogen Peroxide to a solution of Cuprous ions, e.g. Cuprous Chloride, could produce a precipitated mixture of Cu and CuS. In so far as i know this has never been checked.)

It is also known that the tarnish on Silver is Silver Sulfide, not Silver Oxide. Could something similar be happening there? Would carefully cleaned Silver plate sealed in pure Oxygen still tarnish with Silver Sulfide? There may be chemistry of this type all around us at the fringes of molecular and nuclear chemistry which have always been considered, possibly wrongly, as totally different disciplines.

Reference:
1. Sinclair, Dean LeRoy, <u>Oscillators in a Substance Model of Existence: A Physicists' Grail and an Alexander's Sword</u>, (Large-print, underlined.) Sinclair, Dean LeRoy.<u>The Skinny of OiaS</u>: <u>The Small Print Editon of</u> <u>Oscillators in a Substance Model.</u> (The same content as the other book, smaller print, no underlining.) CreateSpace/Amazon, 2015

Final Comments

It was originally intended to add an appendix covering the Four Factor Coding of a portion of the naturally occurring isotopes of the Periodic Chart along with a few examples of the use of the Outer/Inner Levels Model. However, it was realized that probably, this would be better done in another book, more encyclopedic in scope of nuclear reactions.

A Four Factor Charting of the Naturally Occurring Isotopes through the first three periods is included in the books, <u>Oscillators in a Substance Model of Existence</u>, and its smaller print version, <u>The Skinny of Oias</u>,

www.ingramcontent.com/pod-product-compliance
Lightning Source LLC
Chambersburg PA
CBHW081317180526
45170CB00007B/2746